Armor to Venom

Armor to Venom
Animal Defenses

Phyllis J. Perry

A First Book

Franklin Watts
A Division of Grolier Publishing
New York London Hong Kong Sydney
Danbury, Connecticut

For Jay and Randy

Note to Readers: Terms defined in the glossary are **bold** in the text.

Photographs ©: Animals Animals: 17 (Steven Dalton), 14 (Breck P. Kent), 8 (Anup & Manoj Shah); Denver Zoo: 36; Jeff Wines: 24, 51; John G. Shedd Aquarium: 38; Melissa Stewart: 34; Photo Researchers: 13 (Scott Camazine), 16 (Manfred Danegger/OKAPIA), 49 (Jack Daniels), 32 (Nigel Dennis), 31 (Francois Gohier), 53 (M.P. Kahl), 29 (G.C. Kelley), 30 (Tom & Pat Leeson), 50 (C.K. Lorenz), 40 (George Lower/National Audubon Society), 46 (Fred McConnaughey), 15 (Tony Mercieca), 44 (John Mitchell), 27 (Andrew Rakoczy), 35 (Martin Wendler/OKAPIA); Valan Photos: 56 (Jim Merli); Visuals Unlimited: 20 (Wm. C. Jorgensen), 42 (A. Kerstitch), 2 (E.C. Nielsen), 55 (Science VU), 23 (Marty Snyderman), 43 (Justan W. Verderber), 10 (R. Wallace).

Library of Congress Cataloging-in-Publication Data

Perry, Phyllis Jean.
    Armor to venom: animal defenses / Phyllis J. Perry
    p.   cm. — (A First book)
    Includes bibliographical references and index.
    Summary: Describes how animals survive by using their armor, camouflage, horns, stings, and other natural protective devices and strategies.
    ISBN 0-531-20299-2 (lib. bdg.)   0-531-15884-5 (pbk.)
    1. Animal defenses—Juvenile literature.   [1. Animal defenses.]   I. Title.   II. Series.
QL759.47—dc21   96-37289
                                                                                      CIP
                                                                                       AC

© 1997 by Phyllis J. Perry
All rights reserved. Published simultaneously in Canada.
Printed in the United States of America.
1 2 3 4 5 6 7 8 9 10 R 06 05 04 03 02 01 00 99 98 97

Contents

CHAPTER 1

Predators and Prey

9

CHAPTER 2

Avoiding the Enemy

12

CHAPTER 3

Animals With Armor

19

CHAPTER 4

Animals With Weapons

26

CHAPTER 5
Animals With Venom
39

CHAPTER 6
Unusual Animal Defenses
47

GLOSSARY
57

FOR FURTHER READING
59

INDEX
61

Armor to Venom

▲ *A swift impala runs for its life in a frantic atempt to escape the jaws of a hungry cheetah.*

CHAPTER 1

Predators and Prey

A male impala grazes quietly, its long **horns** curved gracefully back above its head. Suddenly its head jerks up, and its ears flicker this way and that. A tiny crackling sound has alerted it to danger.

Only yards away, a crouching cheetah suddenly springs from its hiding place. The swift impala bounds off, zigzagging in a desperate effort to outrun its enemy. The cheetah leaps through the air and knocks the impala to the ground. Without wasting a single moment, the cheetah suffocates the impala with its sharp teeth, and then sits down to enjoy its meal.

There are more than a million different animal **species** on our planet. Most of them serve as food for other animals. In the wild, animals are constantly hunting or being hunted. The hunters are the **predators**, and the hunted are the **prey**.

▲ *This decorator crab adds to its effective camouflage by attaching algae and sea anemones to its body.*

Predators, such as the cheetah, are not always successful. This is because prey animals use a variety of defenses to protect themselves. Some of these defenses help animals avoid their enemies altogether. A decorator crab is often able to hide from its predators. The crab attaches bits of algae, seaweed, or small creatures to its body, so that it blends into its environment. Gulls and other potential predators often pass right by without even seeing the crab.

Other defenses help animals protect themselves when they must face their enemies. When hyenas try to attack zebras, the zebras snap with their teeth and slash with the hooves of their hind legs. Since this type of defense is not enough to discourage hungry lions, zebras resort to another defense **strategy**. When a herd of zebras notices a lion approaching, the zebras run back and forth in a random pattern. The moving mass of stripes confuses the lion, so that it is unable to single out a zebra for attack.

Although prey animals use a variety of defenses, predators do often succeed in killing their prey. Despite this killing, the total population of prey animals usually stays about the same. Prey animals do not disappear altogether because they have higher birth rates than the animals that hunt them. Most prey animals mate earlier and produce more young than their predators. The delicate balance of life is maintained because prey animals are present in much greater numbers than their predators.

CHAPTER 2

Avoiding the Enemy

When you think of animal defenses, you probably think of horns, hooves, sharp teeth, and claws. While these weapons are useful, many animals protect themselves by avoiding their enemies. To save itself, an animal may hide, go underground, or try to make a fast getaway.

Out of Sight, Out of Mind

Some animals have coloring that allows them to blend in with their surroundings. The lichen katydid is almost invisible against a tree trunk, while the green tree toad matches surrounding branches and leaves. The arctic lemming sheds its brown summer fur for a white coat as winter approaches.

Some creatures rely on their spotted or striped coloration for protection. The fur of young deer is dappled with spots that resem-

▲ *Deep in the Everglades of Florida, a green tree toad blends so closely with the its surroundings that it is almost invisible.*

ble sunlight and shadow. When these deer stand still, they blend in with the forest underbrush. The speckled feathers of some birds make them nearly invisible among leaves or tall grass. Some birds also have an eyestripe, which helps conceal their eyes.

▶ *The kallima, or dead-leaf butterfly, lives up to his name.*

Then there are the great pretenders. These include the batfish and the kallima butterfly. Both look just like leaves. Some harmless insects mimic wasps or bees, so that predators will confuse them with stinging insects.

Going Undercover

Some animals protect themselves by living part of their lives underground. A snake called the side-winding adder can sink into the sand in seconds. A trap-door spider lifts the silk-hinged door of its **burrow** and comes out only to catch prey of its own.

European badgers live in underground towns called **setts**. A sett has at least two entrances. A network of tunnels leads to the nursery and sleeping chambers. Badgers sleep during the day and come out at night to hunt for food. When chased by an enemy, a badger will duck down into one of its tunnels and fiercely threaten the predator with its teeth and claws.

If a rabbit is lucky, it can dodge a hungry fox long enough to reach its burrow, dive in, and disappear below the earth. When a prairie dog hears a warning bark from another member of its

▼ *A trap-door spider cautiously waits for prey so close to its burrow that if an enemy approaches, it can quickly disappear back inside.*

15

▲ The amazing earth-moving feet of the mole help it tunnel quickly.

colony, it rushes underground. Moles use their shovel-like front feet to build their burrows. They can dig a 12-foot (3.7-m) tunnel in a single hour. If threatened on one of their rare trips above ground, moles quickly disappear into their maze of branching tunnels.

Beavers and muskrats build lodges with underwater entrances. The dens that otters dig in riverbanks have underwater back doors that act as emergency exits.

Run for Your Life

If hiding and burrowing fail, a threatened animal may try to make a fast getaway. The amazing basilisk lizard drops from a riverside tree, runs

▶ *Dropping from a tree or bush over the water, a basilisk lizard runs a considerable distance on top of the water berfore sinking and then swimming to safety.*

16

on water until it loses speed, and then swims away. A flying squirrel avoids a pursuing marten by gliding from one tree to another.

Using their long, slender, muscular legs, animals like antelope, impala, and gazelles can race like the wind. They often run in a zigzag path, making them harder to catch. Mountain goats and sheep run with surefooted skill among towering cliffs. Kangaroos and rabbits are strong, agile hoppers.

Most birds can fly away from their enemies. A few, like the ostrich and the roadrunner, run to escape from their predators. An adult ostrich can easily outrun a lion. A roadrunner can run as fast as 20 miles (32 km) per hour.

CHAPTER 3

Animals with Armor

You may have seen pictures of gallant knights riding out to the jousting field carrying lances. The knights were well protected in suits of armor. Many animals are similarly protected by tough skin, scales, shells, and armor plates.

You Can't Get Under My Skin

Some animals are protected by their tough skin. The thick skin of the African badger, commonly known as the honey badger, protects it from bee stings as it breaks open hives and eats honey. The skin also protects the badger from the claws and teeth of its predators.

An aardvark goes out at night, sniffing along the ground until it finds a termite mound. Using its sharp claws, the aardvark tears the mound apart and inserts its long, mucus-covered tongue to catch

the termites. The aardvark's thick hide protects it from the sting of termites and ants as it eats.

Snakes have hard, horny scales that protect them from predators. Some snakes even have a clear scale over each eye. These scales protect their eyes from dirt and from being scratched. Crocodiles and alligators have either heavy scales or thick, bony plates.

A shark's skin is covered with toothlike scales called **dermal denticles**. These scales, which are made of the same material as a shark's teeth, help sharks defend themselves against enemies.

The bodies of crabs and lobsters are enclosed in a thick shell called an **exoskeleton**. As a lobster grows, its shell eventually becomes too tight. When this happens, the shell splits across the back and the lobster crawls out, already wearing a new, soft shell. This larger shell quickly hardens and protects the lobster from many of its enemies.

Other sea creatures have shells that act like armor, too. Clams, mussels, scallops, and oysters all have two hinged shells. When they close these shells, powerful muscles help them keep the two halves tightly closed.

The hermit crab protects its soft body by climbing into an empty seashell and carrying it around on its back. When the crab outgrows its protective home, it searches the beach for a new one.

◀ *This lined shore crab has grown too large for its old shell and is molting.*

Small octopuses may climb into large triton shells to hide from predators while they are eating.

The snail carries its shell around with it wherever it goes. When it senses danger, it quickly draws its entire body inside the shell. The hard bottom of its foot acts like a plug, protecting the snail from its enemies. A predator cannot get at the snail unless it can break the snail's shell.

The box turtle also pulls itself into its tough outer shell when predators are nearby. The top part of a turtle's shell is called the **carapace**, and the flat, bottom portion is called the **plastron**. The front and back plates of the plastron are hinged, so that the turtle can easily pull itself completely inside. The dome shape of the carapace makes it especially strong, so most predators cannot break it with their jaws.

The Ultimate Armor

While some animals rely on their tough skin, scales, and shells for protection, others are actually covered in armored plates. A suit of bony plates, welded together and covered in scales, safeguards the backs and heads of armadillos. Only their bellies are unprotected. Flexible bands across the middle of an armadillo's bony exterior act like hinges and allow the animal to bend and twist. When attacked,

▶ *An octopus has found an empty shell in which to hide from its enemies.*

the armadillo rolls itself tight into a ball to protect its stomach from enemies.

The pangolin, which lives in Africa, may grow up to 6 feet (2 m) in length and weigh 60 pounds (27 kg). Like aardvarks, these animals use their long tongues to eat ants and termites. The back of a pangolin is covered with overlapping horny scales that look something like the shingles on a roof. These plates or scales have sharp edges. When a predator approaches, the pangolin swings its tail and drives its knifelike scales into its enemy. Sometimes the pangolin wraps its armored tail around itself to protect its scaleless underside.

Like lobsters and crabs, insects have a tough outer skin called an exoskeleton. This exoskeleton, which is jointed much like a suit of armor, is made of a substance called **chitin**. Chitin cannot grow, so when an insect gets too big for its coat, the old coat splits and the insect crawls out. A new coat of chitin hardens around the insect.

◀ *A banded armadillo is in the process of curling up. When it is in a tight ball, its stomach will be protected from attacking enemies.*

CHAPTER 4

Animals with Weapons

When animals are confronted by enemies or rivals, they may fight to protect themselves and their young. In these battles, animals strike out using **antlers**, horns, or tusks on their heads, powerful kicks with their hooves, or vicious rakings with their sharp claws.

An Array of Antlers

Antlers are made of bone and are covered in a soft, velvety layer of skin. This temporary skin delivers blood, which contains important nutrients, to the rapidly growing bone during the spring and summer. Although animals such as deer and moose may use their antlers to protect themselves or defend their territory against predators, antlers are most important for attracting females and driving off rival males during the mating season.

▲ *This bull elk is scraping away dead skin from its magnificent antlers.*

One of the most beautiful antlered animals is the wapiti or elk. Among elk, large antlers are a sign of strength and power within the herd. During the winter, the bull elk sheds its antlers. Each spring, new antlers begin to grow from two permanent bases called **pedicles**.

In autumn, the antlers stop growing and begin to harden. The skin covering the antlers dries, and bull elk rub their antlers against trees and bushes to scrape away the dead skin.

Moose are fiercely territorial animals and have been known to attack sled dog teams and hikers with their antlers. As with elk, however, antlers are most important during the mating season. If one bull moose is challenged by another, the bulls will battle by clashing antlers.

The antlers of an old bull moose may spread as wide as 6 feet (2 m). Moose antlers also emerge from a pedicle. Like elk, moose drop their antlers in the winter and grow new ones in the spring. A moose's antlers are made of bone and are covered in a layer of skin. Female moose normally do not have headgear.

Like elk and moose, the antlers of male deer are covered in a velvet-like skin during the spring and summer. Deer antlers harden in the autumn and are shed each winter.

Both male and female caribou have antlers. The males have large antlers, which they shed each winter. The cows have smaller antlers, which they keep until spring. During the long, cold winter when males and females are competing for scarce food, the antlered females may chase away the defenseless bulls. This allows pregnant cows to get enough food to support the developing young.

◀ *A greater kudu bull has very long, curled horns.*

Horny Headgear

Some horns are straight spikes. Others, such as those on the greater kudu, are curled. While most horned animals have just one pair of horns, some domestic breeds, like the Jacob's sheep, have two pairs of horns. Even though horns come in a wide variety of shapes and sizes, they are all made of a substance called **keratin**. Your fingernails are also made of keratin.

▲ *During mating season, bighorn sheep rams in the Rocky Mountains fight with one another for mates by butting heads.*

The horn of a bighorn sheep is curled and sweeps backward. It consists of a central core of bone that is permanently attached to the sheep's skull. Each spring, the bony core and keratin covering, grow longer and wider. In winter, the growth slows. Bighorn rams never shed their horns.

Rams with large horns are dominant and lead the others to feeding and mating grounds. The bighorn rams do not have a herd of females like elk. They gather on a breeding ground once a year to choose mates. When one ram challenges another, they crash head-to-head—using their horns for battering, not stabbing. Female bighorns, or ewes, also have headgear, but their horns are much smaller. The

30

ewes sometimes use their horns to protect their young lambs.

Mountain goats live among steep cliffs. Their horns are about 7 inches (18 cm) long, pointed, and ebony in color. At one time, Native Americans living in the northwestern United States collected these horns and carved them into spoons and other utensils.

Like elk and deer, pronghorns shed their headgear each winter. The keratin sheath falls off, but the bony core remains. A new layer of keratin begins to form immediately. By spring, the headgear is larger than it was the previous autumn.

The horns of musk oxen grow outward from bone shields on their foreheads. These horns are not shed each year. The members of a musk oxen herd typically huddle together in a circle with their

▲ When facing an attack, musk oxen form a circle with their horns facing outward toward their enemies.

31

horns facing outward. The young oxen remain at the center of the circle and are protected by the adults. Buffalo herds often gather in a similar fashion. If an intruder approaches, they lower their heads, dash forward, and slash the enemy with their horns.

The Indian rhinoceros has just one "horn," while the black and white rhinos of Africa have two "horns" on their heads. These are not

▼ *A white rhino from Zululand displays its two horns.*

true horns, however. Although they are made of keratin, they do not have a permanent bony core. Rhino "horns" grow directly out of the animals' skin, as do our hair and fingernails. When a rhino feels threatened, it will charge the intruder, gore the animal with its horns, and then trample its enemy.

Hooves, Nails, Claws, and Teeth

Many animals use their hooves to protect themselves. When deer feel threatened, they often rear up on their hind legs, then kick and slash their enemies with their front hooves. Kangaroos use the strong nails at the tips of their feet to slash and rip the skin of their attackers. A kangaroo may also bite its enemy and box with its front feet.

When a raccoon faces an attacker, it uses its teeth to tear and the claws on its hind feet to rip at its enemy. Members of the cat family use their sharp claws to capture prey and to fight with enemies. When cats are not using their claws, they sheath them within their paws.

Tusks are especially long canine teeth. They grow in pairs and extend out of the mouth. Tusks are not used to chew, but may be used to dig. Like horns and antlers, tusks are often used to fight.

To protect its young from a polar bear, an adult walrus may rise to its full height and use all its might to jab its tusks into the bear. A walrus also uses its tusks to display, to fight with other walruses, and to drag itself out of the water and up onto the ice.

▲ *It is a battle of the giants when two elephants fight one another using their long tusks.*

Male Asian elephants and all African elephants have tusks. These tusks, which may grow up to 10-feet (3-m) long and weigh up to 200 pounds (90 kg), are made of ivory and have an enamel tip. The tusks of females are more slender than those of males.

Elephants use their tusks to defend themselves against other elephants and to fight off lions. In defending their calves, African elephants have occasionally been known to charge automobiles and repeatedly stab the cars with their tusks.

Spines and Quills

Although spines and quills are sharp and hard like thin knitting needles, they are nothing more than a special type of hair. The hedgehogs of Africa, Asia, and western Europe and the porcupines of North America are protected from many predators by their prickly pelt.

▲ *A western European hedgehog protects itself by rolling into a ball with its spines standing up straight.*

▶ *Since porcupine quills have barbs, they are hard to remove from an animal that has been foolish enough to attack.*

The body of a hedgehog, which ranges from 5 to 12 inches (13 to 31 cm) in length, is covered with several thousand spines. When frightened, muscles beneath the hedgehog's skin contract, causing the spines to stand up straight. The hedgehog's sharp spines are white, with a brown stripe around the middle and at the tip. When it needs to, the hedgehog can roll itself into a ball so that it is completely protected by its spines.

Porcupines are rather slow, clumsy rodents with spiked quills. These quills are 1 to 3 inches (2.5 to 7.6 cm) long, hollow, and have barbs along the shaft. Once the barbs dig into a predator's flesh, they are very difficult to remove. The quills may eventually work their way deep into an animal's body and pierce its heart, lungs, brain, or liver.

A porcupine's coat consists of a combination of soft guard hairs and about 30,000 quills. Although the quills are soft when the porcupine is first born, they harden in about 30 minutes. When a porcupine encounters a predator such as a coyote, wolf, fisher, or cougar, it turns away from the attacker and raises its quills. If the attacker tries to bite the porcupine, it will get a mouthful of quills. The porcupine may also strike its enemies with its quill-covered tail.

The echidna of Australia, also called a spiny anteater, has quills similar to those of a porcupine. An adult echidna is usually about 24 inches (61 cm) long, including its tail. The echidna captures its dinner by thrusting its mucus-covered tongue into insect nests and swallowing the insects that stick to the mucus. If threatened, it may curl into a spiny ball or dig into the ground, leaving just the tips of its sharp spines exposed. Unlike most of its **mammalian** relatives, young echidnas hatch from eggs.

There are even some fish with spines. One group, called the scorpion fish, includes the beautifully-striped zebra fish. The spines on the back and fins of zebra fish contain a strong poison, which the fish uses to protect itself from predators. The stonefish, the deadliest of all the scorpion fish, has thirteen spines. It lives in the Red Sea, the Indian Ocean, and along the northwestern coast of Australia. These fish often lie at the bottom of tide pools. The hatpin sea urchin, which may be found in tide pools or coral reefs, has long, poisonous spines that are as sharp as needles.

▲ *The spines along the back of this dragon stonefish contain a strong poison.*

Some caterpillars are covered with spiny hairs. At the base of each of these hairs is a poisonous gland. The lasiocampid moth caterpillar has patches of spiny hairs beneath a fold of skin on its back. If attacked, this caterpillar arches its back and tries to drive its spiny hairs into its predator.

CHAPTER 5

Animals with Venom

In the previous chapter, you read about some animals with poison-filled spines. These are not the only animals that defend themselves with poisonous chemicals. **Venom** can also be delivered to an attacker with tentacles, fangs, or stingers. A variety of creatures, large and small, use venom. While some use their poison to protect themselves, others use it to kill prey.

All of these animals make their venom in special glands. Some venoms dissolve the victim's body tissues, and others make it difficult for the victim to breathe. Still others stop the victim's heart. Some types of venom paralyze the victim temporarily, while others can kill in a matter of minutes.

A Portuguese man-of-war, a colony of tiny sea creatures, lives in warm ocean waters throughout the world. When it floats on the surface, it looks something like a balloon with lots of streamers

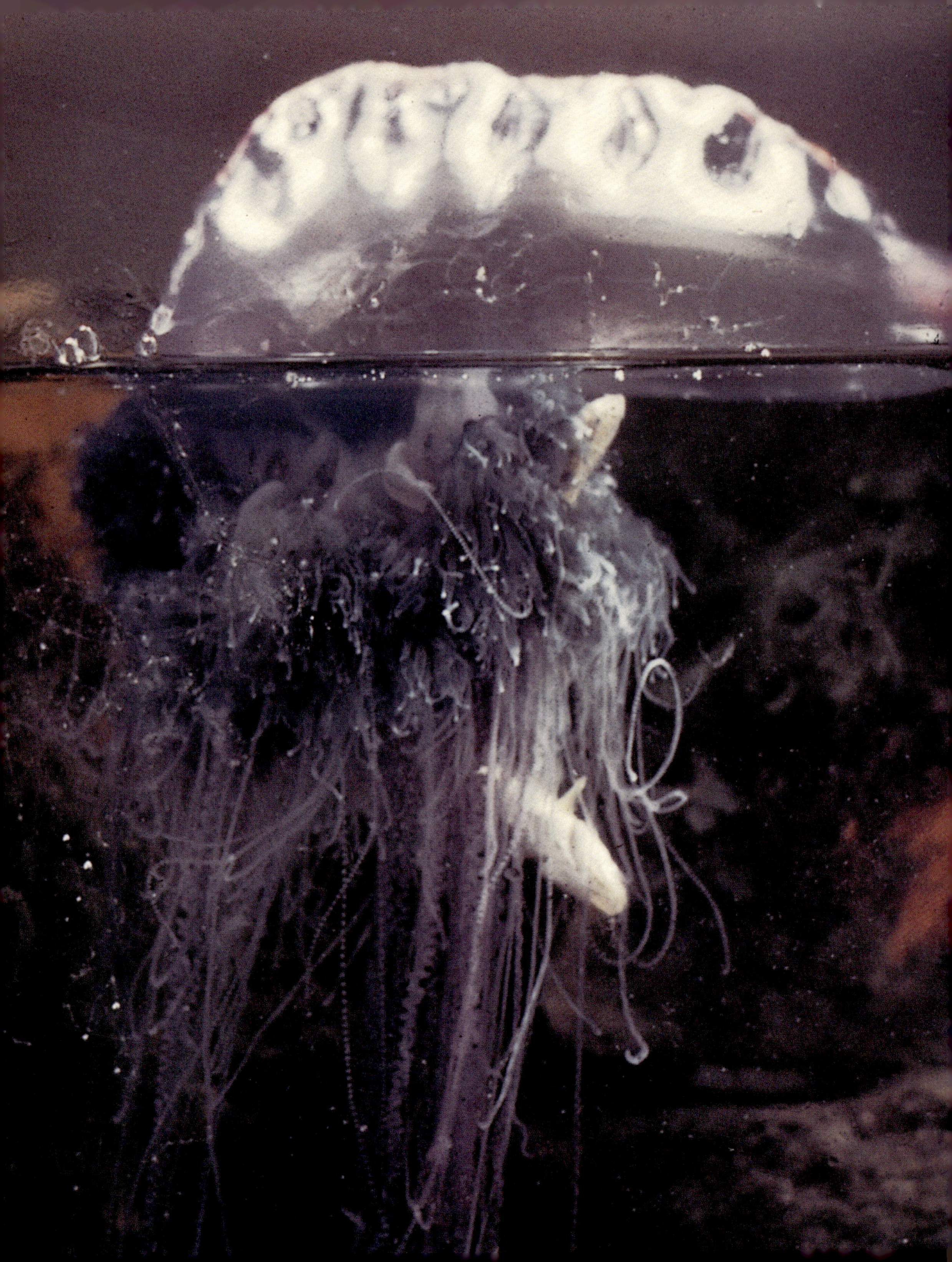

hanging down from it. These "streamers," which are called tentacles, may be up to 50 feet (15 m) long. The tentacles are covered with bead-like stinging cells called **nematocysts**.

When a fish, shrimp, or other small ocean animal swims close to a Portuguese man-of-war, it may become tangled in the tentacles. Each time the fish touches a tentacle, every bead on that tentacle releases a poisonous barb. These barbs, which look something like harpoons, quickly kill the victim, providing a tasty meal for the Portuguese man-of-war.

One sea animal, the *Glaucus atlanticus,* can immobilize these stinging cells and feed on the tentacles of the Portuguese man-of war. The glaucus is then able to use the stinging cells it has eaten in its own defense system.

Beautiful sea anemones are also common in ocean waters all over the world. Like the Portuguese man-of-war, sea anemones poison their victims and attackers with nematocysts. When the sea anemone poisons a prey animal, it uses its tentacles to pull the prey into its mouth. The brightly-colored clown fish may swim in and out of an anemone without being harmed because it has a mucous coating that protects it from the poison.

◀ *This small fish is being stung to death by poisonous barbs released from the tentacles of a Portuguese man-of-war.*

▲ *A Mexican cone shell mollusk searches for fish. A poisonous barb shot out of the open end of its shell will stop the heart of its prey.*

The cone shell mollusk lives in tropical waters. It can use either armor or venom to protect itself. Like its relatives the clam and snail, the cone shell mollusk is enclosed in a hard shell. It also has teeth that act something like a harpoon gun. When a cone shell mollusk comes into contact with a predator, it sends a poison-filled, harpoonlike barb out of the open end of its shell. The powerful venom, which is more deadly than a rattlesnake's, stops the heart of the victim almost immediately.

The venomous stingray, which has a poison-filled, spearlike structure in its tail, lives in shallow temperate and tropical ocean waters. Stingrays usually lie on the ocean floor, half-buried in the sand. When disturbed, a stingray swings its tail in all directions. The poison-filled spear often cuts, and may even remain stuck in, the intruder. When a stingray loses its poisonous spear, it grows a new one.

The giant water bug may be only 2 inches (5 cm) long, but it can kill a large bullfrog. The water bug uses its front legs to grab its enemy as well as its prey. The water bug then attacks its victim with its venom-filled beak.

When a stranger approaches a beehive, the guard bees become very agitated. As a result, their barbed stingers, which resemble sharp, hollow needles, fill with venom. As you may know, when a bee stings an enemy, it releases the venom into its victim. Besides containing poisonous chemicals, the venom also contains strong-smelling chemicals that alert other bees of the danger. While a bumblebee can sting its victim many times, a female honeybee can sting only once.

◀ *A giant water bug grabs its dinner and then kills it using a venom-filled beak.*

▲ *The scorpion curls its tail forward and jabs a poisonous stinger into its victim.*

There are more than 600 species of scorpions. All of them live in tropical environments, and each one has a stinger and venom-producing glands on the tip of its tail. When it is attacking a smaller animal, a scorpion holds its victim in its strong pincers, brings its tail forward, and jabs its stinger into the victim. The scorpion then uses its strong ripping claws to tear the victim into pieces.

Most spiders are at least mildly venomous. To poison a victim, a spider must bite the prey with its fangs—hollow tubes in the front part of its body. Two of the more dangerous spiders are the female black widow and the brown recluse, both of which are found in the United States. Although spider "bites" can make people very sick, actual deaths are very rare.

Some, but not all, snakes are venomous. Common poisonous snakes include the rattlesnake, the boomslang, the cobra, the coral snake, and the cottonmouth water moccasin. The poison is introduced into another animal through special grooved teeth or fangs, which are hollow. These snakes have special poison-producing glands at the base of their fangs. The spitting cobra of Africa sprays venom into the eyes of its enemies.

Shrews are small, furry animals that live in many parts of the world. The short-tailed shrew has a venomous bite. When the short-tailed shrew bites another animal, venom from the glands in the shrew's jaws numbs the victim.

Some animals, such as the puffer fish, are poisonous only if eaten. The puffer fish, which lives in tropical regions of the ocean and large rivers such as the Congo and the Nile, tries to discourage its enemies by puffing itself up so it looks much bigger than it really is. When this fish is eaten, poison in its skin and internal organs destroys the nerve cells of its predators. In a few cases, people have died from eating these fish.

▲ *The porcupine fish puffs itself up to look bigger than it really is. If eaten, its poison can make people sick or even kill them.*

Some toads also have poisonous fluids in their skin. When the toad is squeezed in a predator's jaws, the poison flows to the surface of the toad's skin. Marine toads are extremely poisonous and have occasionally killed dogs that swallowed them.

Newts, which are related to frogs and toads, may also be poisonous. The crested newt which lives in Europe, Asia, North Africa, and North America has venom in its back and tail. The skin, blood, and muscles of the California newt contain venom. Even its eggs are poisonous.

CHAPTER 6

Unusual Animal Defenses

Avoiding enemies, relying on protective armor, and using weapons such as horns or poison are the most familiar animal defenses. There are, however, a number of other less common animal defenses.

Playing Dead or Injured

In a tight situation, some animals may decide to play dead. This strategy often works because many predators will only eat a fresh kill. The most famous pretender is the common opossum, which is found in North, Central, and South America. When an enemy approaches, the opossum falls and goes limp. Its lips curl back and its heartbeat slows. After a predator passes by, the opossum gets up and runs in the opposite direction.

The hognose snake, which lives in sandy regions of the eastern, western, and southern United States, is harmless and has no fangs or poison. It usually feeds on toads, frogs, and lizards. When this snake comes face-to-face with its own predators, it puts on quite a show. First, it pretends to be a venomous snake. It rears up as if to strike, hisses, and coils its tail. If this doesn't frighten off the attacker, the hognose will fall to the ground and appear to have convulsions. Finally it rolls onto its back and hangs its mouth open so that it looks dead.

When a mother osprey sounds a note of warning, her fledglings will partially extend their wings, droop their heads over the edge of the nest, and remain completely still. Any passing predator will surely think they are dead. When the danger has passed, their mother will call them a second time.

Some birds, such as the grouse, instinctively act as if they have a broken wing or leg and lead a predator away from a nest full of young birds. The predator follows, thinking the injured bird will be an easy catch. When the predator has been lured far enough away, the grouse flies up into a tree. She flies back to her chicks or eggs a few moments later. This behavior is called a distraction display.

▶ *A hognose snake, which is not poisonous, may try to frighten off an attacker by rearing up as if to strike. It may also roll onto its back and pretend to be dead.*

49

Look at Me!

Instead of hiding, some animals advertise themselves. Their brilliant coloring or markings make them stand out from their surroundings. The bold color schemes of poisonous lizards and frogs let enemies know that eating them would be a mistake. These colors scream: "Leave me alone if you value your life!"

Most warning patterns are made up of combinations of red, orange, yellow, black, and white. The poisonous gila monster and the beaded lizard of the southern United States have bands of black and yellow or orange markings, as does the blue-tongued skink of Australia.

▶ *A blotched blue-tongued skink opens its mouth wide in a colorful and threatening display.*

▼ *This poisonous gila monster makes its home in the Sonora Desert of Arizona.*

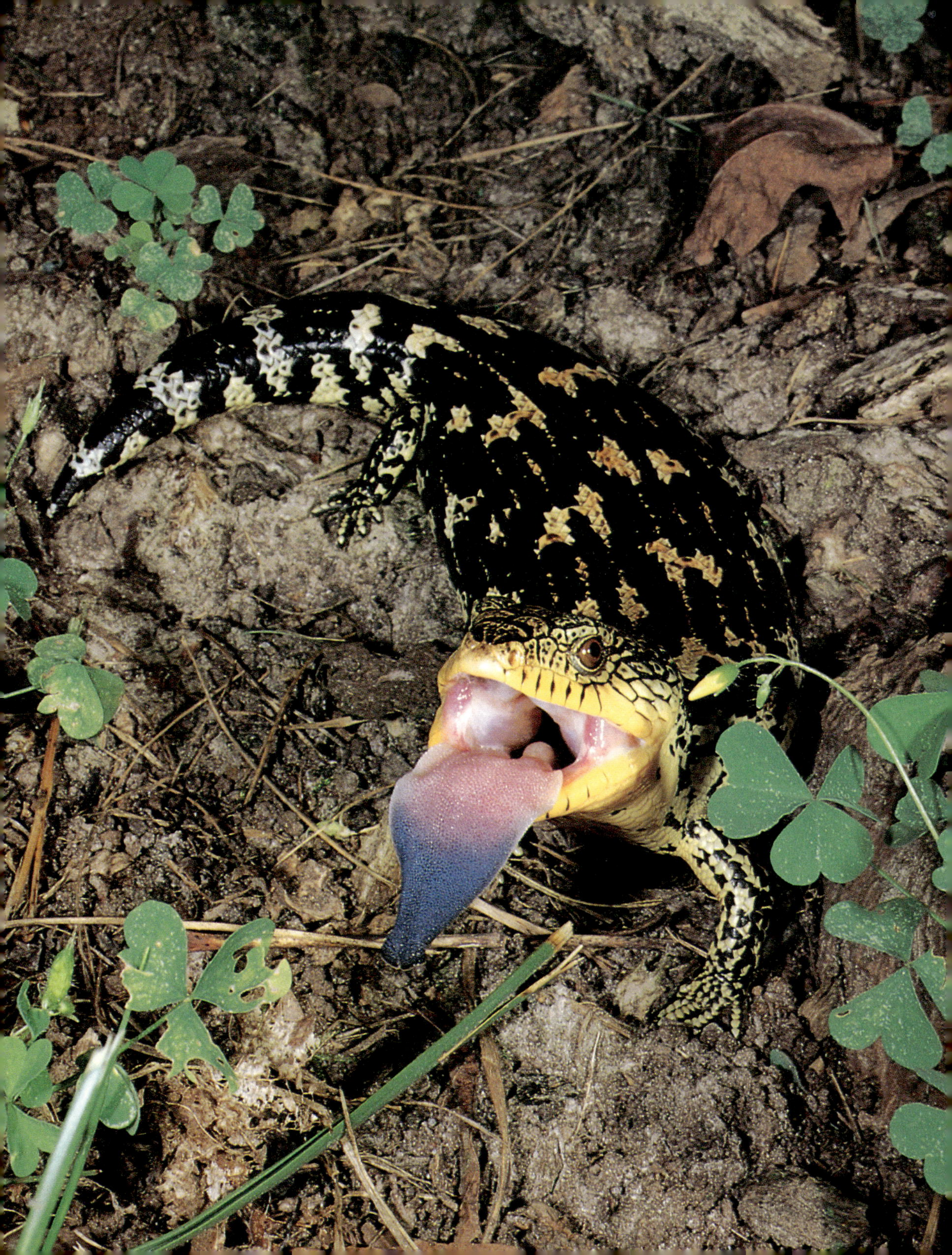

Predators associate a skunk's white stripe with its noxious spray. Skunks are most active at night, but even then their white stripes can still be seen. When a skunk is threatened, it turns its back on its predator and sprays it. A skunk can hit a target that is 10 feet (3 m) away. Besides the dreadful smell, the spray stings the predator's eyes and may even cause temporary blindness.

There's Safety in Numbers

A troop of baboons may include as many as 150 individuals. When baboons travel from one area to another, they move in set formations. Young males lead the way. They are followed by smaller females and older juveniles. Nursing mothers and babies are escorted by the largest and strongest males.

When baboons stop to rest, the strong males and nursing mothers sit in the center of the group. Younger males guard the edge of the troop. A few may also be posted as lookouts on rocks or in trees. If an enemy approaches the troop, females and youngsters go ahead and the males drop back, baring their teeth and making threatening gestures.

Social insects such as ants and termites also live in groups. Each individual in the group is assigned a specific task. Some gather food, while others act as soldiers or guards.

▶ *The younger and smaller baboons remain near the center of a traveling group, protected by males who guard the edges of the troop.*

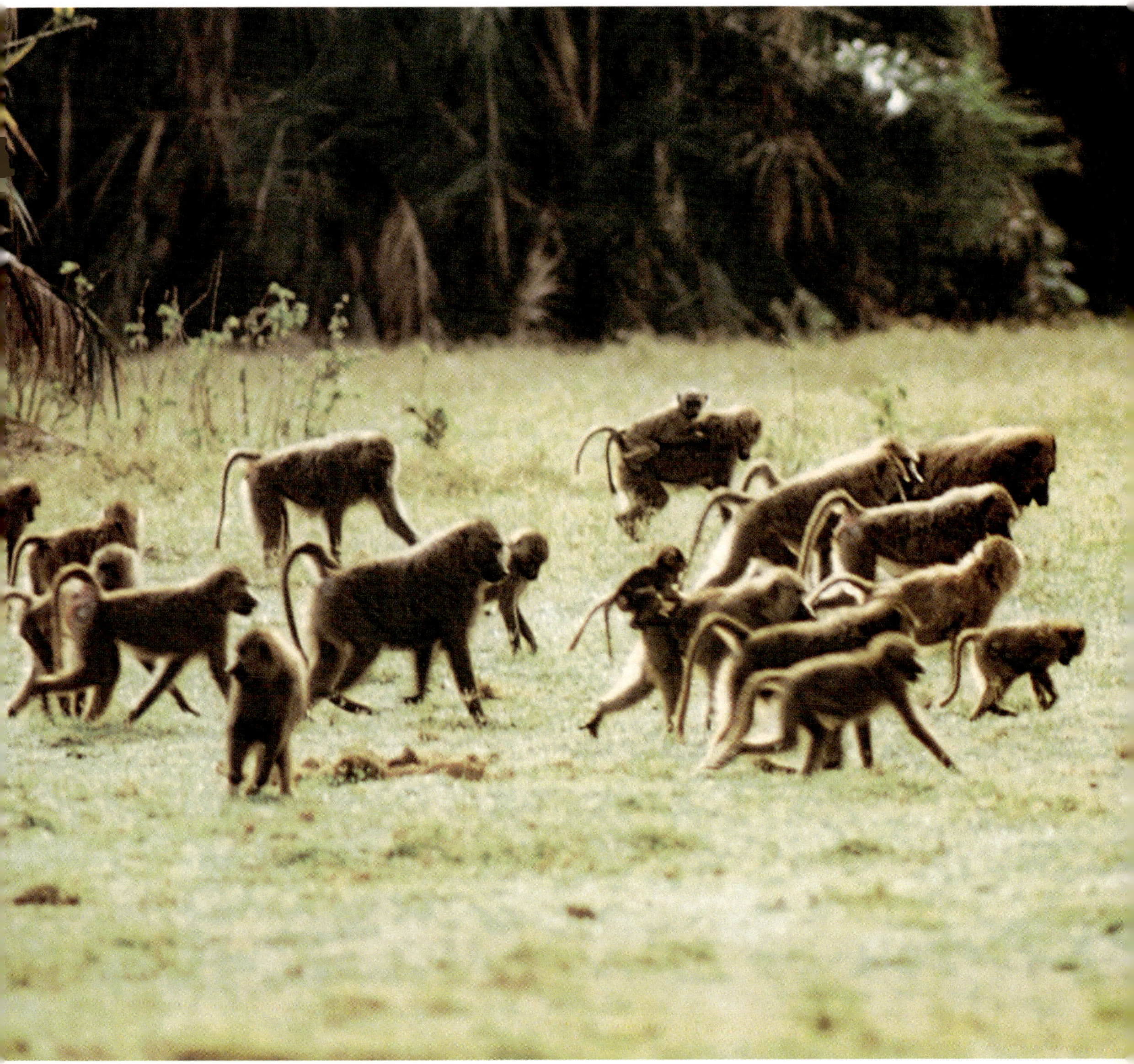

Protective Friends

Sometimes one species of animal can convince a larger, stronger one to protect it. Aphids, a common pest in any rose garden, are the preferred food of one of our favorite insects—the ladybug. Because aphids are easy prey for ladybugs, they need a special defense against their enemy. Ants are the answer.

Ants use their strong, biting jaws to protect the tiny, defenseless aphids from ladybugs and other predators because the aphids produce a sweet, sticky substance (honeydew) that the ants enjoy eating. Aphid colonies that are protected by ants actually reproduce more rapidly so they can turn out more honeydew. If you find a branch loaded with aphids, chances are you'll also find ants. They will be crawling about among the aphids.

Other Tricks of the Trade

When an octopus feels threatened, it can discharge an ink-like chemical and then quickly swim away. In most cases, the predator, such as a moray eel, will be distracted by the ink smoke screen and lose its chance to capture the octopus. Cuttlefish and squids can also discharge an inky chemical.

▶ *Defenseless aphids produce a sweet substance to feed the ants that protect them from ladybugs and other enemies.*

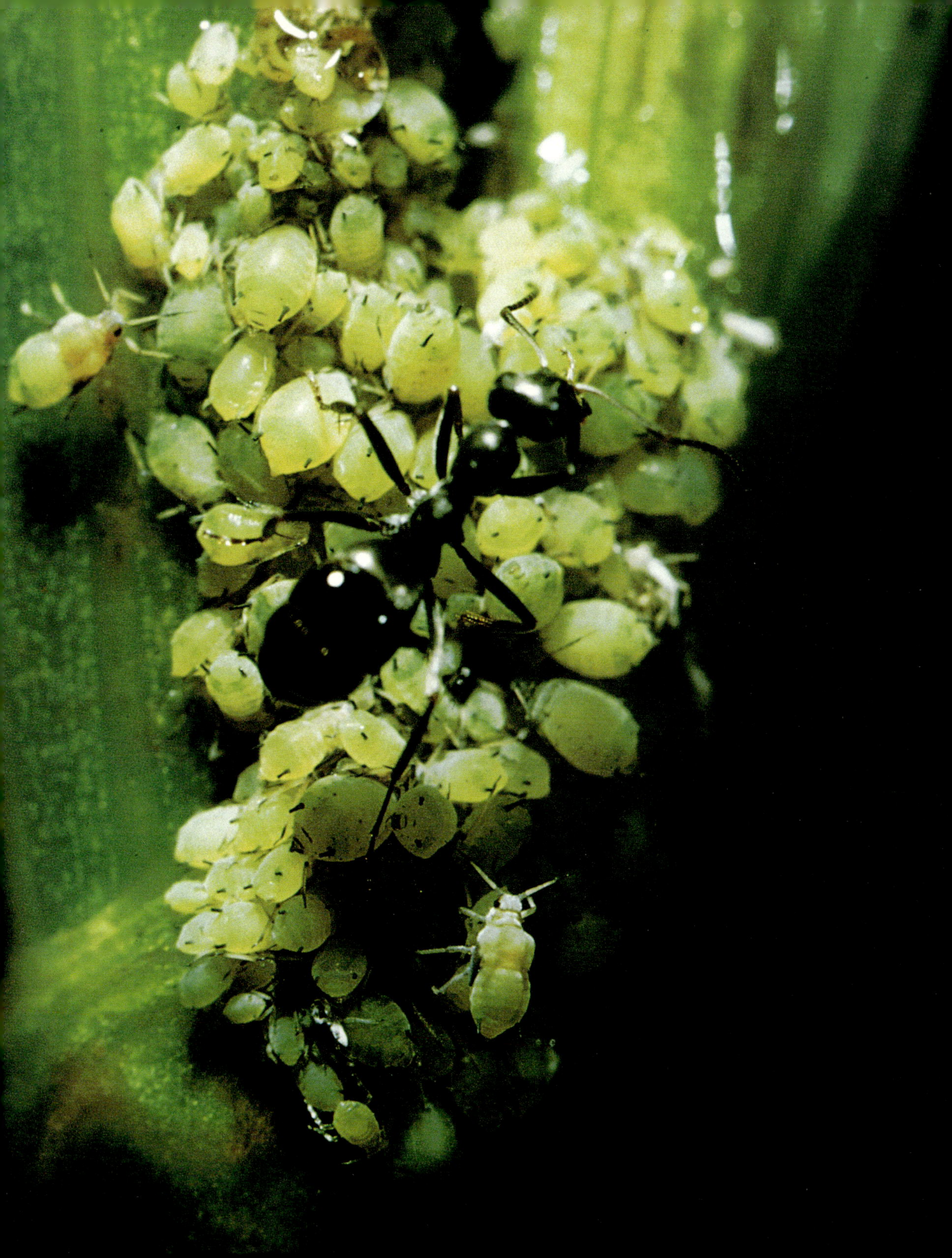

Some lizards and salamanders have a very unusual defense. They do not have poisonous skin like their relatives the newts and toads, nor do they make use of armor or venom. They do not have weapons, and they do not play dead. Instead, they make use of their unique breakaway tails. If a predator grabs one of these animals by the tail, the animal will release its tail and run away.

Like the spear of the stingray, the tail of the lizard or salamander will eventually grow back.

▼ *This lizard, a midnight swift, would rather give up its tail than its life. Eventually the breakaway tail will grow back.*

Glossary

antlers a type of animal headgear that is shed each year. It is made of bone and covered with skin while growing.

burrow a passage dug in the ground by an animal.

carapace the top half of a turtle's shell.

chitin the protein that makes up the exoskeltons of insects and our fingernails.

dermal denticle a pointed, toothlike object that sticks out from the skin of an animal, such as a shark,

exoskeleton the hard, protective layer that surrounds the body tissues of animals such as crabs, lobsters, and insects.

horn permanent headgear made up of a bone base and a keratin covering.

keratin a tough protein fiber from which fingernails, claws, hooves, and horns are made.

mammalian refering to a mammal. The mammals are a group of animals that includes elephants, dogs, cats, horses, kangaroos, whales, bats, seals, dolphins, humans, etc.

nematocyst the beadlike stinging cells that cover the tentacles of jellyfish and the bodies of anemones.

pedicle the permanent bone base of antlers.

plastron the bottom half of a turtle's shell.

predator an animal that kills other animals for food.

prey an animal that is hunted for food.

sett a underground community of animals such as badgers.

species a group of organisms that produce viable offspring when they mate.

strategy an adaptation that helps a plant or animal to live a longer life and reproduce more efficiently.

venom a poisonous fluid produced in the bodies of some animals and introduced into the bodies of others by stings or bites.

For Further Reading

Arnold, Caroline. *Elephant.* New York: Morrow Junior Books, 1993.

Bennett, Paul. *Escaping from Enemies.* New York: Thomson Learning, 1995.

Duprez, Martine. *Animals in Disguise.* Watertown, MA: Charlesbridge Publishing, 1994.

Machotka, Hana. *Outstanding Outsides.* New York: Morrow Junior Books, 1993.

Perry, Phyllis J. *Hide and Seek: Creatures in Camouflage.* Danbury, CT: Franklin Watts, 1997.

Simon, Semour. *The Optical Illusion Book.* New York: W. Morrow and Company, 1984.

Sowler, Sandie. *Amazing Armored Animals.* New York: Alfred A. Knopf, 1992.

Zingg, Eduard. *Rhino!* Edina, MN: Abdo and Daughters, 1993.

Index

Antlers, 26-28, *27*
Aphids, 54, *55*
Armadillo, 22, *24,* 25
Armor, 19. *See also* Chitin; Scales; Shells; Skin; Plates

Baboons, 52, *53*
Bighorn sheep, 30-31, *30*
Birth rates, 11
Burrow, 14-16, *16*
Butterfly, kallima, 14, *14*

Camouflage, 10, *10. See also* Color
Carapace, 22
Chitin, 25
Claws, 19, 33

Crab, 10, *10, 20,* 21
Color
   blending in, 12-14, *13,* 14
   standing out, 50-52, *50, 51*

Death. *See* Playing dead
Defenses, 10. *See also* Hiding: Playing dead; Running; Weapons
   faking injuries, 48
Dens, 16
Dermal denticles, 21
Distraction display, 48

Elephants, 34-35, *34*
Elk, 27-28, *27,* 31
Exoskeleton, *20,* 21, 25

Eyes, 13, 21

Fish, 14, 37, *38,* 41, 45, *46*
Fly away, 18

Gliding, 18
Greater kudu, 29, *29*
Groups, safety in, 52, *53*

Hair. *See* Quills; Spines
Hedgehog, 35-36, *35*
Hiding, 10, *10. See also*
    Underground
  and color, 12-14, *13, 14*
Hooves, 11, 33
Hoppers, 18
Horns, 8, 9, 29-33, *29, 30, 31,*
    32
  pronghorns, 31

Impala, *8,* 9, 18
Injuries, faking, 48

Keratin, 29, 30, 31, 33

Lizards, 16-18, *17,* 50, *51,* 56,
    56

Mating season, 26, 28, 30, *30*
Mimics, 14
Mole, 16, *16*
Mucus, 19, 21, 37, 41
Musk oxen, 31-32, *31*

Nails, 33
Nematocysts, 41

Octopus, 22, *23*

Pedicles, 27, 28
Plastron, 22
Plates, 21, 22, *24,* 25
Playing dead, 47-48, *49*
Poison, 39
  bees, 43
  caterpillars, 38
  and color, 50-52, *50, 51*
  cone shell mollusk, 42, *42*
  fish, 37, 45, *46*

Glaucus atlanticus, 41
hatpin sea urchin, 37
newts, 46
Portuguese man-of-war, 39-41, *40*
scorpions, 44, *44*
sea anemones, 41
shrew, 45
snakes, 45
spiders, 45
stingray, 42-43
toads, 46
water bug, 43, *43*
Porcupine, 36-37, *36*
Portuguese man-of-war, 39-41, *40*
Predators, 9-11
Pretenders, 14, *14*
   faking injury, 48
   playing dead, 47-48, *49*
Prey, 9-11, *10*
Protection, obtaining, 54

Quills, 35, 36-37, *36*
Rhinoceros, 32-33, *32*
Running, 8, 9, 11, 16-18, *17*

Scales, 21, 22, 25
Scorpion, 44, *44*
Setts, 15
Shells, 20, 21-22, *23*, 42, *42*
Skin, 19, 21
   covering antlers, 26, *27*, 28
Smoke screens, inky, 54
Snakes, 45, 48, *49*
Species, 9
Spiders, 14, *15*, 45
Spines, 35-38, *35, 38*
Strategy, 11

Tails, 25, 37, 42-43
   breakaway, 56, *56*
Teeth, 9, 11, 33, 52. *See also* Tusks

Tentacles, *40,* 41
Toads, 12, *13,* 46
Tongue, 19, 21, 25, 37
Tunnels, 15-16, *16*
Tusks, 2, 33-35, *34*

Underground, 14-15, *15, 16*
Underwater entrances, 16

Venom. *See* Poison

Water bug, 43, *43*
Weapons, 26. *See also* Claws; Horns; Poison; Teeth
    antlers, 26-28, *27*
    hooves, nails, 33
    spines and quills, 35-38, *35, 36, 38*

ABOUT THE AUTHOR

Phyllis J. Perry has worked as an elementary school teacher and principal and has written two dozen books for teachers and young people. Her most recent books for Franklin Watts are *The Crocodilians: Reminders of the Age of Dinosaurs*, *The Snow Cats*, and *Hide and Seek: Creatures in Camouflage*. She received her doctorate in Curriculum and Instruction from the University of Colorado, where she supervises student teachers. Dr. Perry lives with her husband, David, in Boulder, Colorado.